Degradation of African Socio-cultural Values in the Age of Technology : A Critical Engagement with Okyere Manu, Ubuntu and Communal Ethics

Akom Stephan Ngam

Published by Smartworld Publishers, 2026.

DEGRADATION OF AFRICAN SOCIO-CULTURAL VALUES IN THE AGE OF TECHNOLOGY : A CRITICAL ENGAGEMENT WITH OKYERE MANU, UBUNTU AND COMMUNAL ETHICS

First edition. September 2, 2026.

ISBN: 979-8233757938

Written by Akom Stephan Ngam.

Table of Contents

To my parents, Mr & Mrs Ngam.

This research has undergone double-blind peer review by the editorial board of the Smartworld Journal for Research and Innovation [SWJRI]. The review process assessed originality, methodology, and contribution to knowledge.

ACKNOWLEDGMENTS

One of the pillars of formation is education. Thus, I wish to begin by thanking Almighty God for the graces bestowed upon me throughout the course of this work. My endeavours could have been futile without his blessings, protection and inspiration.

Secondly, I express my profound gratitude to my supervisor, Rev. Dr. Remi Prospero Fonka (PhD); whose inspiration at the beginning of my philosophical journey and keen attention in this work has been so valuable.

Equally, I extend heartfelt thanks to the CATUC academic family. In a special way, I appreciate my lecturers, learning infrastructure, and research opportunities that enhanced my learning. The lecturers of the department of philosophy, nurtured philosophical units and morals in me.

My Mill Hill family is appreciated for the enormous privileges granted to me; especially the opportunity to study at CATUC Bamenda. I am sincerely indebted to the formation staff, Fr. Andrew Mitema M.H.M. and Fr. Ignatius Shing M.H.M., whose profound desire for my success was demonstrated through their spiritual, financial, moral, and material support.

To my classmates, especially those with me in formation, I extend my thanks for standing by me during difficult moments and for their constant companionship. Finally, I owe immeasurable gratitude to my beloved parents, whose sacrifices, prayers, and unwavering support both moral and material have been the foundation of my journey. Their love and encouragement have sustained me throughout this endeavor.

AKOM SEPHAN NGAM

JUNE 2026

Abstract

African societies today grapple with the degradation of socio-cultural values under the pressures of rapid technological advancement. This dissertation critically engages Okyere-Manu's ethical reflections on degrading African socio-cultural values, situating her concerns within the broader discourse on *Ubuntu* and communal ethics. She argues that technology, often shaped by Western individualistic paradigms, risks marginalizing indigenous knowledge systems and weakening communal solidarity if adopted uncritically. Using a philosophical and analytical methodology, this study examines her perspectives on ethical frameworks, *Ubuntu's* relational morality, and the limitations of technology in safeguarding African socio-cultural values. It further evaluates her call for responsible ethics, highlighting tensions between socio-cultural preservation and economic growth, as well as the risks of overstating values at the expense of diversity. Applying these insights to contemporary Cameroonian society, the dissertation underscores the relevance of *Ubuntu* and ethical responsibility in navigating technological transformation. The findings affirm the imperative of integrating African ethical values into policy, education, and innovation, ensuring that technological progress promotes human dignity, communal well-being, and sustainable development. In doing so, the work highlights Okyere-Manu's contribution and advances a framework for re-reading African ethics and socio-cultural values in the digital age.

Key Terms: *African socio-cultural values, technological advancement, communal ethics, indigenous knowledge systems, Western individualism, communal solidarity, ethical frameworks, relational morality, responsible ethics, human dignity, sustainable development.*

General Introduction

African societies today face a profound challenge; the degradation of socio-cultural values under the weight of rapid technological advancement. Practices that once defined communal life, oral tradition, respect for elders, solidarity expressed in rituals, and ethics of responsibility, are increasingly neglected by digital platforms and Western-driven innovations; according to John S. Mbiti.[1] This erosion of cultural identity is not simply a sentimental lament; it is a pressing ethical and socio-cultural concern. This threatens the very foundations of African communal existence and moral responsibility.

Okyere-Manu[2], in her ethical and socio-cultural concerns, consistently raises issues needing the intersection of African values to streamline technological innovations. She argues that technology is never neutral but culturally embedded; often carrying Western individualistic paradigms that clash with African communal values. Through the lens of *Ubuntu*[3], she insists that Africans must interrogate technological imports, whether they align with values of dignity, solidarity, and responsibility. Her works highlight the dangers of uncritical adoption of technology innovations; warning that without ethical safeguards, Africa risks losing its cultural identity, and weakening socio-ecological resilience.[4] Okyere-Manu's ethical concerns, thus, provide a vital framework for rethinking how African societies can humanize technology and reclaim urgency in the digital age.

The central problem addressed in this research is growing disjunction between African socio-cultural values and innovations in technological systems imposed upon them. While technological innovation promises economic growth, improved healthcare, and enhances education, it simultaneously undermines communal solidarity, compromises values of indigenous languages, and disrupts pertinent traditional practices. The challenge lies in reconciling technological progress with preservation of essential socio-cultural values. Thus, how can African societies embrace technological innovations without compromising their rich moral heritage? How can policymakers, educators, and communities ensure that technology serves humanity and its dignity without commodifying it? These questions form the core of our investigation (problem statement); seeking to critically assess Okyere-Manu's ethical concerns, confronting the degradation of African socio-cultural values.

Furthermore, the research Methodology employed is a philosophical and analytical approach. It draws on textual analysis of Okyere-Manu's works related African philosophical literature; situating her ethical concerns within broader African philosophical debates on technology, ethics, and cultural identity,[5] as projected by Kwasi Wiredu. The methodology therefore, engages in a critical and interpretive style; combining conceptual analysis with contextual application to Cameroon particularly and African realities in general.

Our work is divided into three chapters. Chapter one explores Okyere-Manu's ethical perspectives in relation to African socio-cultural values; emphasizing on *Ubuntu* communal values and the limitations of uncontrolled technological innovations. Chapter two provides an appraisal of her reflections; prescribing responsible ethics, assessing power dynamics, and highlighting the necessity of cultural preservation. Lastly, chapter three projects the relevance of the study to contemporary Cameroon; highlighting, the significance of ethical responsibility in navigating technological innovation and transformation. The ends concludes with a general conclusion; synthesizing findings and

proposing a bibliography for further research.

Chapter One

An Understanding Of Okyere-Manu's Ethical Perspectives In Relation To African Socio-Cultural Values

Okyere-Manu emphasizes the importance of considering African values and ethics into technological development to prevent socio-cultural degradation.[6] Her work highlights the need for African perspective to ensure that technology aligns with African cultural context. In this chapter, focus is on Okyere-Manu's conceptualization of the ethical perspectives apropos African values. This establishes a solid foundation justifying the relevance of views in facing degrading African socio-cultural values. She examines in the following themes: background influence and ethical context, *Ubuntu* communal philosophy and socio-cultural degradation linked to technological progress.

1.1. Background Influence and ethical contexts

OKYERE-MANU'S BACKGROUND influence and ethical context informs her approach to analyzing the impact of technology on African socio-cultural values. She argues that these values, especially ethical concerns, must be streamline to control technological development; ensuring that emerging technologies do not just compromise, but promote and preserve African socio-cultural values and practices.[7] This foundational stance is explored through two subthemes: ethical influence and African analytical framework, and the central argument of African cultural values.

1.1.1. Ethical Influence and African Analytical Framework

OKYERE-MANU ADVOCATES for an ethical lens through which technology should be viewed in relation to African socio-cultural values. She contends that most technologies are designed within cultural frameworks of non-African societies; often reflecting foreign values. Therefore, she insists it is essential to evaluate technological advancements alongside western sanctioned ethically and socio-cultural norms, to ensure accommodative relevance and integration into African life.[8] Okyere-Manu extent the argument contending that:

> Equally, African Indigenous Knowledge Systems have existed long before the advent of Western Knowledge Systems, and they embody practices, skills, and epistemologies that are deeply rooted in communal ethics and cultural continuity.[9]

She warns that without an ethical framework that pays attention to African contexts, technology risks become an alien imposition, rather than an empowering tool. Thus, the analytical framework she proposes insists on balancing imported technological norms and practices with indigenous ethical resources, to safeguarding African identity and ensuring resilience.[10]

1.1.2. Central Argument of African Cultural Values

IN ADDITION, EVERY culture possesses core values defining its identity. Okyere-Manu asserts that the dominant producers of technological innovations are unfortunately rooted in Western cultural backgrounds.[11] Consequently, she thinks these technologies are often developed without consideration for African cultural values, which are distinct and foundational to African societies. This oversight leads to the erosion and compromise of African values.[12] Given this concern, Okyere-Manu turns to the *Ubuntu* communal philosophy as a vital ethical resource for confronting technological challenges in African contexts.

1.2. Manu And *Ubuntu* Communal Philosophy

THE *Ubuntu* philosophy is central to Okyere-Manu's thoughts; emphasizing African relational and moral theories in global discussions especially confronting technological development. She further contends that *Ubuntu* provides a framework for understanding ethical implications and responsibility in the face of technological advancement of African societies; promoting responsible technological development. Here, she insists that the community is important in the same way as the individual; since whatsoever happens to the individuals affects the whole community.[13]

More so, she asserts that "the marginalization of African Indigenous Knowledge Systems in favor of Western epistemologies has created a disjuncture between African cultural values and the technologies imposed upon them"[14] by invoking *Ubuntu*, she is asserting for a corrective approach that re-centers African communal ethics, ensuring that technological innovations do not erode but rather reinforce the values of solidarity, reciprocity, and respect for life. In this way, Africa will reclaim its cultural identity and strengthen socio-ecological resilience, ensuring that technological innovations serve communal well-being rather than undermine it.[15] This theme is explored through two dimensions: *Ubuntu* and relational morality vis-à-vis global technology, and *Ubuntu* and responsible ethical attitudes.

1.2.1. Ubuntu and Relational Morality Vis-a Vis Global Technology

ACCORDINGLY, OKYERE-Manu sees the sense of community as a central socio-cultural and ethical African value. Other values, among many for her include the ethics of *Ubuntu* which encompasses the sense of hospitality, communal responsibility, and the sense of good relations with others; respecting institution, elders, and authority. In her estimation, global technological innovations directly affect ethical concerns; ranging from the disappearance of traditional jobs, irresponsible physical exertions causing psychological harm to humanity and environment, general dehumanization, and social rupture of societal interpersonal relationships.[16]

Furthermore, Mark Findlay cites Okyere-Manu as contending that no artificial intelligent system can be considered as a general-purpose intelligent agent, performing well in a wide variety of environments, without compromising proper ability of human intelligence. This prompts him to highlight how negative currents of social media are increasingly disrupting African communal values of parenting. Equally, he concludes pointing out how the advancement of sex robots clashes with African belief in natural fertility and family propagation.[17]

1.2.2. Ubuntu and Responsible Ethical Attitudes

ACCORDING TO MANU, cultivating behaviours promoting interconnectedness and community is an aspect of responsible ethical attitudes. She asserts that most artificial intelligent applications interrupt authentic African interpersonal affiliations; specifically fashioned to oust African context contextual relevance. Out rightly, she insinuate that such artificial intelligent systems clashes with African traditional values and realities; instrumentalizing human beings enthroning machines, becoming real danger to African humano-ethical responsibility.[18]

In another dimension, Okyere-Manu proposes the following questions: How can a society that has been guided by the tenets of communitarianism survive amid artificial intelligent system advancements, developed in an individualist context? What are long-term implications for such cultures?[19] Manu in these formulated interrogations regrets the disappearance of family time, once a space for transferring knowledge, values, and life skills; is gradually disappearing, increasingly attention is diverted from humans to global media. The consequence of this is diminishing understanding of African realities, socio-cultural values, and traditions.

1.3. Socio-Cultural Degradation And Technological Advancements

AFTER ESTABLISHING ethical foundations and *Ubuntu* framework, Okyere-Manu turns to broader issues of socio-cultural degradation, caused by technological advancement. She warns that technology, if not aligned with African values and practices, easily erodes communal life and cultural identity. Out-rightly for her, a compound understanding of the complex relationship between technology and socio-cultural values in African contexts is necessary.[20] As face-to-face interactions diminish, this becomes technology's pitfall, confronting socio-cultural values; compelling her to urge African scholar's indigenes to reflect on the interface of ethics, socio-cultural values, and technological development.[21] In light of these limitations, a culturally grounded understanding of technological advancement becomes not only necessary but urgent.

1.3.1. Limitations of Technology Confronting Socio-Cultural Values

OKYERE-MANU OUTLINES several limitations of technology affecting African socio-cultural values: erosion of traditional practices, digital tools replacing oral storytelling, communal rituals, and other aspects of indigenous knowledge systems. She propounds that global technological systems shaped by Western epistemologies marginalize African languages and communal values.[22] She warns that privacy norms designed for individualistic societies feel alien in African communal cultures. By invoking *Ubuntu*, she insists that Africa must reclaim its identity and ensure technology reinforces solidarity, reciprocity, and respect for life.[23] Furthermore, OKyere-Manu warns that technological determinism risks displacing core African values; exhibiting individualistic and utilitarian paradigms:

> It must be noted that technology is constantly evolving with new and improved innovations which continue to affect some aspects of human, life. Evidently, the dominant producers of these innovative and technological developments are from the western cultural background and values, suggesting that the design and development has been advanced without Africa and specifically African cultural values that distinguish them distinctly from other societies in mind.[24]

Thus, the limitations of technology are not only technical, but also deeply ethical and cultural. This leads naturally to the next concern: how can African societies understand and respond to technological advancement in a way that preserves their values?

1.3.2. Necessity Understanding Technological Advancement: Socio-Cultural Implications

IN RESPONSE TO AFOREMENTIONED limitations, Okyere-Manu calls for a critical and culturally grounded understanding of technological advancement. She emphasizes that while innovations in artificial intelligence, biotechnology, and educational technology offer transformative potential, they also risk deepening inequalities and eroding indigenous values if adopted uncritically. This prompts her to advocate for an African-centered lens that prioritizes communal well-being, ethical reflection, and cultural preservation.[25]

Okyere-Manu further explores how technologies shape human identity and social relationships. Accordingly, she warns that the integration of psychtechnology and robotics into everyday life may alter traditional African conceptions of personhood, community, and moral responsibility. For instance, in her estimation, increasing reliance on digital platforms for education and communication could weaken face-to-face communal bonds; so central to African ethics such as *Ubuntu*; urging scholars and policymakers to interrogate not only the utility of these tools but also their deeper cultural consequences.[26]

Subsequently, Okyere-manu advocates for a proactive ethical framework that empowers African societies to shape technological trajectories rather than merely consuming them. This for her involves fostering local innovation, embedding cultural values into design processes, and ensuring that technologies serve human dignity and social justice. Her reflections further resonate with broader calls for decolonizing technology[27] and affirming African agency in the digital age.[28] This becomes a vital contribution to the discourse on how Africa can navigate the promises and dangers of technological change with wisdom and cultural integrity.

1.4. Policy and Practical Implications

OKYERE-MANU EMPHASIZES the significance of integrating African ethical values into policymaking processes; arguing that policies in disregard of indigenous ethical frameworks, risks being ineffective and culturally alienating.[29] For policymakers, this for her means recognizing that African communities possess rich moral traditions, which fundamentally guide social behavior, communal responsibility, and justice. By aligning policies with aforementioned values, governments according to Okyere-Manu foster initiatives that resonate more deeply with people's lived realities; thereby improving acceptance and successful implementation of developmental programmes.[30]

Central to this integration in Okyere-Manu's thoughts, is the concept of *Ubuntu*; which she described not as an abstract ideal, but a living ethical framework governing individual and community relations. She further asserts that *Ubuntu* embodies the principle of shared humanity and mutual responsibility; guiding not only interpersonal relationships but also informing governance structures that prioritize collective well-being over individualism. She insists that such inclusive governance, rooted in socio-cultural values are essential to rethinking developmental policies in African contexts.[31] Okyere-Manu turns to how policymakers can deliberately instill African values within developmental frameworks; extending immediate concerns of governance, and demonstrating the widening of developmental significance of ethical values into economic, political, and environmental spheres.

1.4.1. Policy Makers' Integration of African Ethical Values

BUILDING ON THIS ETHICAL foundation, Okyere-Manu explores practical implications of incorporating African ethical values into developmental works. She demonstrates how ethical values like *Ubuntu*, respect for elders, and communal harmony, act as catalysts for social cohesion; considered indispensable for sustainable development.[32] For example, in her research, she highlights that the respect accorded to elders as custodians of community wisdom plays a critical role in mediating resource-based conflicts, ensuring that technological development projects align with local norms and avoid external impositions.[33]

1.4.2. Significance of African Ethical Values to Development

EXTENDING BEYOND SOCIAL and governance dimensions, Okyere-Manu underscores broader significance of African ethical values in shaping economic, political, and environmental progress. Accordingly, these values in her estimation, offer a moral compass that guides equitable resource distribution and environmental stewardship; both considered as critical for inclusive growth. The African worldview, she insists, regards nature not as a simple resource stockpile, but as a partner deserving respect and balance. Her conclusion is the ethos of harmony with nature in African ethics, provides critical insights for sustainable development policies, counteracting exploitative practices that contribute adversely to climate change.[34]

Furthermore, by grounding governance in shared cultural norms and ethical accountability, Okyere-Manu estimate that African core values contribute significantly to institutional legitimacy. Additionally, she thinks they also promote distributive justice; challenging inequalities often perpetuated by Western developmental paradigms.[35] Subsequently, Okyere-Manu calls for a deliberate and conscious incorporation of ethical perspectives into technological developmental frameworks; aimed at authentic and sustainable progress of African societies.[36]

Okyere-Manu's ethical perspectives highlight the enduring relevance of African cultural values, particularly *Ubuntu*, as a moral compass in navigating the challenges of modern technological advancement. By grounding ethics in communal responsibility and rational morality, she underscores the need to balance innovation with preservation of socio-cultural identity. Her analysis reveals that while technology offers risks eroding the very values sustaining African communities, it becomes an imperative to integrate ethical safeguards, rooted in indigenous frameworks. Ultimately, this constitutes a call for policymakers and educators to harness African ethical values; not merely cultural heritage, but vital resources for sustainable development, guiding societies toward responsible technological explorations, communal solidarity, and holistic human flourishing. All these facets provide an opportunity to highlight salient points for an appraisal of Okyere-Manu's thoughts, as unveiled in the next chapter.

Chapter Two

An Appraisal Of Okyere-Manu's Ethical Perspectives Confronting Degrading African Socio-Cultural Values

Philosophizing requires the critical development of thought, in order to ensure continuity. In this regard, Okyere-Manu's ethical reflections provide a vital framework for interrogating the degradation of African socio-cultural values in the face of technological advancement; arguing that African ethics, particularly *Ubuntu*, must be emphasized as a guiding principle to contemporary African society.[37] Okyere-Manu emphasizes that technology easily undermines communal solidarity and cultural identity when divorced from indigenous ethical contexts.[38] She situates African values in particular, not as static traditions, but dynamic resources capable of confronting socio-cultural erosion, and degradation.

Inevitably, aforementioned perspectives enable this chapter underscore the necessity of integrating African ethical values and frameworks into technological design, policy, and education; thereby, as Okyere-Manu vindicates, ensuring that progress does not ensue at the expense of African socio-cultural resilience and moral responsibility.[39] The appraisal, in this chapter, centers discussions on Okyere-Manu's reflections on responsible ethics and technological development, where she insists that African values must guide innovation and adaptation in contemporary society;[40] *ubuntu* communal philosophy and power dynamics, tension between cultural preservation and economic growth and exaggeration of cultural values preservation compromising diversity.

2.1. Responsible Ethics And Technological Development

OKYERE-MANU, IN HER reflections on responsible ethics and technological development, emphasizes that African values outstandingly guide how societies adopt and adapt to technology. She insists that ethics should not be an afterthought, but a foundational principle in innovation.[41] For Africa, this requires prioritizing communal well-being, dignity, and relationality over profit-driven or individualistic models.[42] Yamikani Ndasauka in like manner, criticizes Western models of technology as exploitative and calls for African traditions of communal and ecological ethics; insisting that technology must harmonize with people and environment,[43] reinforcing Okyere-Manu's call for relationality. Gyekye defends a moderate communitarianism, balancing individual rights with communal obligations;[44] this, supports Okyere-Manu's insistence that Africa should resist purely profit-driven models and instead prioritizes collective dignity and well-being.

Equally, Okyere-Manu makes a call for embedding indigenous moral frameworks into technological development; to ensuring that innovation aligns with African ways of life.[45] Accordingly, Yamikani Ndasauka in accord cautions that the accelerating wave of techno-social change requires careful reflection on how they interact with time-honoured African values.[46] This perspective perfectly complements Okyere-Manu's; stressing the fact that technological progress must be critically examined through the lens of African ethics, of responsible care and integral explorations.

Furthermore, the above voices affirm the fact that responsible technological development is not merely about profit making but about safeguarding cultural identity and sustainability. This position is further emphasized by Bukunmi Deborah Ajitoni; highlighting *Ubuntu* as a dynamic philosophy of interconnectedness. He justifies the fact that *Ubuntu* is not only a cultural heritage; but also a practical ethical framework for governance, social justice, and conflict resolution, underscoring its relevance in guiding technological adoption.[47] Correspondingly, Raphael Mbendera shows that implanting *Ubuntu* into contemporary developmental strategies is essential for restoring solidarity and resilience in African societies.[48] Mutually, above scholarly contributions justify Okyere-Manu's insistence on responsible ethics in technological development; forming part of a broader intellectual movement. Okyere-Manu's vision, further introduces possibilities of the notion of safeguard ethics; positioning African values particularly *Ubuntu* as corrective force against socio-cultural erosion brought about by uncritical technological adoption. We are going to look at, safeguard ethics reversing socio-cultural degradation and African ethical values as streamlining technological excesses.

2.1.1. Safeguard of Ethics Reversing Socio-Cultural Degradation

OKYERE-MANU UNDERSCORES that African ethics, particularly *Ubuntu*, must act as a corrective force against the erosion of communal life, caused by uncritical technological adoption. She warns that shifting intimate sexual relations from human beings to machines largely driven by Western innovation risks commodifying intimacy, weakening communal bonds, and undermining dignity; in African thought, intimacy is communal, spiritual, and tied to fertility and the continuity of life.[49]

Therefore, by safeguarding ethics, Okyere-Manu's vision could possibly and fittingly reserve socio-cultural degradation of African values; that technology serves human flourishing rather than eroding cultural integrity. Ndasauka, once more in support of Okyere-Manu, argues that technological adoption in Africa must be decolonized and re-rooted in communal values; preventing Western technological development from eroding cherished cultural integrity.[50] Kwame Gyekye's defense of moderate communitarianism further vindicates Okyere-Manu's position, by insisting that personhood as supportive of values, is realized through community.[51] Collectively, these voices reinforce Okyere-Manu's safeguard ethics, emphasizing that technology must serve human flourishing by reconsidering intimacy, dignity, and communal life within indigenous African values.[52]

2.1.2. African Ethical Values as Streamlining Technological Excesses

OKYERE-MANU'S SAFEGUARD of ethics establishes foundations for understanding how African values keep a check on technological excesses; insisting that adoption of technological advancements without ethical grounding, risks commodifying intimacy, erosion of communal solidarity, and undermining dignity.[53] For her, *Ubuntu* functions as a corrective principle, re-centering innovation within communal responsibility and relational dignity. This insight traces the path for dialogue with other African philosophers; correspondingly maintaining the fact that technology must be tied to the moral grammar of community. Kwame Gyekye's defense of moderate communitarianism, for instance, reinforces Okyere-Manu's entitlement showing that personhood is realized through community; thereby curbing technologies that elevate autonomy at the expense of solidarity.[54]

Constructing on Okyere-Manu's framework, other philosophers extend the regulative vision of African ethics. Mogobe Ramose similarly interprets *Ubuntu* as a jurisprudential principle; arguing that justice and governance must be grounded in interconnectedness.[55] This complements Okyere-Manu's insistence that technology must be judged on its grounds of sustaining communal harmony, rather than disrupting it. Kwasi Wiredu's call for conceptual decolonization further strengthens Okyere-Manu's position; warning that uncritical adoption of Western technological paradigms risks epistemic dependency and cultural erosion.[56] In this way, Okyere-Manu's safeguard ethics serve as the pivot around which these philosophical perspectives converge. This underscores African ethical values acting as normative filters, streamlining technological excesses by demanding innovation check by communal accountability, human dignity, and the sanctity of life. Philosophically, this collective stance situates African ethics as a horizon of restraint and orientation; ensuring that technological progress serve human flourishing, rather than becoming an autonomous force of alienation.

2.2. *Ubuntu* communal philosophy and power dynamics

UBUNTU, as a communal philosophy, resonates across metaphysics, ethics, political philosophy, and epistemology; affirming that human existence is relational: "I am because we are." In concrete terms, *Ubuntu* rejects atomistic conceptions of the self, situating personhood within a web of communal interdependence. In agreement, James Ogude notes that "the full development of personhood comes with shared identity and the idea that an individual's humanity is fostered in a network of relationships."[57] Furthermore, Benda Hofmeyr underscores that *Ubuntu's* metaphysical relationality analyses atomistic selfhood; emphasizing that personhood is constituted through affective relations rather than autonomous rationality.[58]

In the domain of philosophy of technology, Yamikani Ndasauka's complements Okyere-Manu's stance, stressing the fact that *Ubuntu* resists Western instrumental rationality; offering instead a humanistic framework for collective empowerment.[59] From the perspective of social and political philosophy, *Ubuntu* just as buttressed by Okyere-Manu and others, insists that power must be accountable to the flourishing of all; meanwhile, applied ethics provides practical tools for interrogating bioethics, gender justice, and technological development. Taken together, these contributions demonstrate the fact that *Ubuntu* is not merely a cultural ideal; but a critical philosophical paradigm that integrates metaphysical, ethical, political, and epistemological dimensions, aimed at, interrogating power dynamics while affirming communal flourishing, in technological usage. Constructing, on *Ubuntu's* highlighting of communal flourishing conditions the exercise of power; becoming a tool for exploring how this paradigm confronts contemporary challenges, particularly the tension between cultural preservation and economic growth.

2.3. tension between cultural preservation and economic growth

UBUNTU philosophy emphasizes communal ethic of solidarity, mutuality, and cultural rootedness; often confronting modern economic imperatives. Okyere-Manu argues that technological and economic imports risk destabilizing indigenous practices, unless filtered through communal ethics.[60] This resonates Kwame Gyekye's moderate communitarianism; insisting that cultural identity deserves preservation, even as societies modernize, rejecting both rigid traditionalism and uncritical globalization.[61] From the standpoint of political philosophy, Amartya Sen's *Development as Freedom* analyses growth measured solely by gross development product; reiterating the fact that development must expand substantive freedoms which include cultural participation.[62] Epistemologically, Paulin Hountondji's evaluation of ethnophilosophy warns against freezing culture as static; urging instead a dynamic and, critical appropriation of tradition coexisting with modernity.[63] Thus, Okyere-Manu's concerns align with a wider range of African and global philosophical discourses; economic growth corresponding with cultural preservation, else progress becomes an alienation.

Inevitably and out rightly, *Ubuntu* offers a reconciliatory framework for cultural and economic harmonization. Okyere-Manu and sympathizers argue that the *Ubuntu* principle, provides a dynamic ethic for sustainable development; insisting that growth must sustain communal identity and ethical reciprocity. Similarly, comparative philosophy enriches this appraisal; with John Mbiti's reflections on African religiosity emphasizing that economic growth that is divorced from cultural values, risks nursing numerous disadvantages.[64] Meanwhile Martha Nussbaum's capabilities approach reiterates the fact that development must enhance human dignity, which includes cultural expression.[65] Thus, collectively, these perspectives show that the tension between cultural preservation and economic growth is not a binary opposition, but a philosophical challenge that requires balancing; *Ubuntu* serving as a paradigm for negotiating modernity without erasing communal identity. Explicitly, this avails the opportunity of socio-cultural value preservation as priority, and the dangers of understanding technological advancement; evident in Manu's thought.

The prioritization of socio-cultural values in developmental discourses reflects philosophical insistence on the need for economic growth to be made ethically accountable to communal identity; projected earlier by Okyere-Manu. Within the *Ubuntu* framework, solidarity, mutuality, and rootedness are not peripheral but foundational to sustainable progress. Rita Kiki Edozie argues in similar likes that *Ubuntu*-inspired economic humanism resists alienation by embedding growth within communal ethics.[66] Temitope Fagunwa reinforces this view, calling for *Ubuntu's* revitalization as a revolutionary ideology that safeguards African identity against globalization's homogenizing pressures.[67] In like manner as Okyere-Manu and relevant authours, Bukunmi Deborah Ajitoni emphasizes that collective identity and social harmony must remain central to governance and justice; situating cultural preservation as a precondition for authentic development.[68]

Furthermore, classical African and global philosophers converge on the aforementioned priority. For instance, Gyekye's moderate communitarianism rejects both rigid traditionalism and uncritical modernization; insisting that cultural identity must guide transformation.[69] This resonates Okyere-Manu's insistence that African socio-cultural values, especially *Ubuntu*, must shape technological adoptions; so that innovation strengthens rather than erodes communal life. Similarly, Paulin Hountondji cautions against hardening culture; urging dynamic appropriation of tradition to meet modern challenges;[70] a position Okyere-Manu echoes by cautioning that technology be critically interrogated, to avoid becoming an alien imposition. Mbiti corroborates, and highlights the spiritual dimension of African communal life, warning that economic growth that is divorced from cultural values risks spiritual failure;[71] a concern Okyere-Manu shares in her criticisms of Western individualistic paradigms. Martha Nussbaum adds that dignity requires freedom to sustain cultural identity, not merely material prosperity;[72] a parallel to Okyere-Manu's call for responsible ethics that safeguards human dignity and communal well-being. Hence, socio-cultural value preservation is not merely a sentimental retreat, but a philosophical and ethical priority; ensuring that modernization develops rather than erase cultural values.

In another dimension, the discourse on technological development reveals both promise and peril. While technology is often celebrated as a driver of modernization and economic growth, its uncritical adoption destabilizes indigenous practices and erodes communal identity. Okyere-Manu cautions that technological imports, if not filtered through communal ethics, risk alienating societies from their original cultural foundations.[73] This danger lies not in technology itself, but in epistemological frameworks treating it as neutral or universally beneficial; ignoring socio-cultural contexts. Kwame Gyekye underscores the fact that modernization must be critically appropriated; warning against blind assimilation of technological practices underming cultural identity.[74] Okyere-Manu complements this by stressing that transformation must be critically appropriated through communal ethics; ensuring that technological practices do not fracture African moral foundations.[75]

Paulin Hountondji similarly analyses the tendency of freezing culture as static; arguing that technology must be engaged dynamically, lest it becomes a tool of dependency and intellectual colonization.[76] Okyere-Manu adds that without ethical accountability, technological dependency risks becoming a new form of bioethical colonization, wherein African societies consume innovation without shaping its moral trajectory. Amartya Sen adds that development measured solely through technologically advancement lenses' growth, neglects substantive freedom; including cultural participation and ethical agency.[77]

In another dimension, Mbiti warns that technological progress that is divorced from spiritual and communal values, risks producing existential failures.[78] Martha Nussbaum reinforces this by insisting that technology must enhance human dignity, not reducing individuals to mere consumers of innovation.[79] Contemporary voices such as those of Rita Kiki Edozie highlight the fact that technological development, if within *Ubuntu* prescription, can be reconciled with communal ethics; but when pursued uncritically, risks perpetuating inequality and cultural erosion.[80]

Thus, the danger of technological development lies not only in innovation itself, but rather in its misapprehension as culturally neutral; lacking from ethical accountability to socio-cultural preservation. Subsequently, technology becomes a vehicle of alienation rather than empowerment. Okyere-Manu's intervention tackles this point; insisting that African communal ethics must serve as an evaluative lens through which technological imports are judged. In perils of danger, it becomes equally important to recognize that excessive emphasis on cultural preservation equally generates new challenges; particularly when it compromises diversity and openness to change.

2.4. Exaggerated Value Preservation: Diversity Compromise

OKYERE-MANU CAUTIONS that technological advancement erodes African socio-cultural values and risks cultural homogenization. However, this perspective underestimates various ways in which technology also preserves and amplifies indigenous traditions. Far from eroding identity, digital platforms equally enable African communities' documentation of long oral traditions, archive endangered languages, and share cultural practices globally. Ovett Nwosimiri emphasizes that technology, when critically engaged, serves as a tool for cultural resilience, allowing African societies to adapt while maintaining ethical frameworks. Thus, rather than eroding values, he insinuates that technology strengthens them; by ensuring their survival in new formats. [81]

Thus, technology has become a vehicle for reclaiming African identity in the global digital era. Mohamed Saliou Camara emphasizes that Africa's cultural diversity must be actively represented in digital spaces; with technology providing the very tools of achieving this. By leveraging digital media, African communities according to him, counter misrepresentation, assert agency, and redefine their place in global cultures.[82] Okyere-Manu's framework reinforces this by demanding that representation in digital spaces face accountable evaluation in relation to African communal ethics; so that cultural diversity is not only displayed but also morally sustained. In this sense, technological advancement does not erode socio-cultural values, but rather empowers Africans to preserve their heritage, diversifying cultural expression, and resisting homogenization. Okyere-Manu's concern is valid; but unfortunately, she overlooks the transformative potentials of technology as a force for cultural preservation and ethical renewal.

This chapter resounding up, affirms that the central issue is not technology itself, but rather the framework of its understanding and application. Okyere-Manu's insistence on communal ethics provides a vital evaluative lens, ensuring that African societies while engaging in technological development learn to implore ways that protect cultural identity yet remaining open to transformation. This balance between caution and possibilities of accommodation, sets the stage for chapter three; exploring the pertinence or relevance of these ethical concerns to contemporary Cameroon. It is at this juncture that the negotiating between tradition, modernity, and globalization, continues to shape communal identity and socio-cultural practices.

Chapter Three

Relevance Of Manus' Ethical Concerns To Contemporary Cameroon Society

Okyere-Manu's ethical perspectives, especially emphasis on *Ubuntu* communal philosophy and responsible technological development, find direct resonance in Cameroon; particularly and specifically within the Kom community of the Northwest Region. Kom society, with its rich traditions of kinship, solidarity, and collective ethical responsibility, exemplifies the *Ubuntu* dictum of "I am because we are. These communal values, expressed through oral traditions and ritual practices, demonstrate the fact that personhood is relational; echoing Okyere-Manu's insistence that ethical frameworks must safeguard socio-cultural values against erosion.[83] This chapter unfolds in four interconnected movements. First, is the redirection of emerging technologies to align with communal values. Second, an examination of technological transformation and its socio-cultural consequences is necessary. Third, the articulation of a framework for ethical responsibility rooted in African traditions, connects to the last item, dealing with the affirmation of *Ubuntu* communal philosophy, as the guiding principle for promoting the common good in Cameroon.

3.1. Redirection of Emerging Technologies

OKYERE-MANU EMPHASIZES that the ethical redirection of emerging technologies is the first necessary step in safeguarding communal identity. Equally, she asserts that innovations such as mobile phones, biotechnology, and artificial intelligence must be critically appropriated rather than passively absorbed; lest they erode African traditions and communal solidarity.[84] Her approach is deeply rooted in "*Ubuntu* philosophy," which insists on communal flourishing; something which technological advancements undermine. This resonates Gyekye's notion of "critical appropriation;" regarding communities as selectively adapting modern innovations to strengthen cultural identity, rather than weaken it.[85] In Kom, Cameroon, where kinship and solidarity remain the bedrock of social life, Okyere-Manu's framework becomes practical; availing opportunities of cooperate system sustaining communal and cultural values to moderate mobile communication, agricultural biotechnology, and digital progress.

In another dimention, Okyere-Manu's concerns align with broader philosophical debates on technology and ethics. Paulin Hountondji's analysis of uncritical assimilation of Western paradigms likewise underscores the need for African agencies to be prudent with technological adoption.[86] Rita Kiki Edozie affirms, asserting that when framed within *Ubuntu* perspectives, technological progress reconciles innovation with justice and inclusivity; ensuring that modernization does not become cultural alienation.[87] In this context, applied ethics becomes crucial; agricultural technologies alongside cultural values should strengthen cooperative farming, while digital platforms endeavour to integrate indigenous knowledge alongside global science. Thus, redirecting emerging technologies in Cameroon is not about rejecting modernization but embedding it within African ethical frameworks, which safeguard identity, solidarity, and sustainable development. If the ethical redirection of emerging technologies safeguards communal identity, then the next challenge lies in understanding how these very innovations reshape socio-cultural fabrics of communities; transforming traditions, values, and modes of solidarity in profound ways.

3.2. Technological Transformation and Socio-Cultural Change

OKYERE-MANU CAUTIONS that technological transformation in Africa, while bringing undeniable benefits, must be ethically redirected to safeguard communal identity.[88] Equally, a particular concern she raises is the replacement of face-to-face interaction by digital communication; noticeably undermining embodied unity and oral traditions, which have historically sustained African communities.[89] The Kom tribe in Cameroon vividly illustrates this tension; here, young people embrace digital tools, while elders lament the decline of oral tradition, cooperative farming, and ritual practices that once defined communal bonds. For Okyere-Manu, technological transformation must therefore be critically managed, ensuring that socio-cultural values remain integral to Cameroon particularly Kom development.

Interestingly, African philosophers in connection reinforce above mentioned ethical concerns with practical solutions. For example, Mbiti insists that African identity is rooted in communality, seen in the famous dictum "I am because we are;" highlighting the fact that technological innovation must serve communal life rather than replace it.[90] This suggests that digital tools just as Okyere-Manu insists, should be re-embedded into cultural practices, (recording folktales or organizing collective rituals online), so that unity is strengthened rather than lost in society. In Kom (Cameroon) just as elsewhere in the Africa, insistence on ethical responsibility by upholding socio-cultural values, acts as a gadfly to technological innovations; avoiding unwanton value degrading and promoting cultural preservation.

3.3. Framework for Understanding Ethical Responsibility

OKYERE-MANU'S INSISTENCE that technological transformation must be ethically redirected is relevant to contemporary Cameroon. She emphasizes that responsibility must be collective; rooted in harmony and justice, and safeguarding the cultural heritage while redirecting modernization.[91] This resonates socio-cultural values of the Kom tribe, highlighted in oral traditions, historically embodying communal responsibility. Nevertheless, modernization and digital culture increasingly weaken these values; creating risks of technology serving only individual interests. Okyere-Manu's recommendation of building an ethical framework of responsibility, offers Cameroon a means of ensuring that progress (technical innovation) strengthens rather than erodes communal identity.[92]

In connection, Jerome M. Tosam highlights how the advent of modernity in Kom has greatly challenged agrarian agriculture and environmental ethics; illustrating how communal farming and ecological solidarity are continuosely endangered by globalization and uncritical adoption of technology.[93] Equally, Walter Gam Nkwi complements in agreement with Okyere-Manu; introducing the Kom concept of modernity or innovation (kfaang). This for Nkwi demonstrates that technological advancement must be negotiated within communal structures of kinship and solidarity.[94] In addition, he insinuates that the Kom society has historically embraced change as a collective project; ensuring that innovations strengthen rather than fragment cultural identity.[95] His ethnographic work shows that modernization, when harmonized with communal validated practices, sustains *Ubuntu's* perception of reciprocity and mutual care. Alongside Okyere-Manu and Tosam, Nkwi's perspective underlines the fact that Cameroon's modernization requires a deliberate ethical framework; one that bridges the gap of technological transformation with cultural preservation and collective responsibility. Yet, no framework of ethical responsibility can be complete without a unifying principle. This is where *Ubuntu* principal becomes indispensable; it transforms ethical responsibility from theory into lived practice, promoting the common good and ensuring that technological progress harmonizes with solidarity, reciprocity, and justice

3.4. *Ubuntu* Communal Philosophy as Promoting Common Good

UBUNTU generally in Africa, as esteemed by Okyere-Manu, provides the unifying principle that becomes relevant to Cameroon. In Kom, an expansion of this is not abstract but lived through cooperative farming, kinship obligations, and ritual gatherings that embody unity and reciprocity. By embedding *Ubuntu* into technological and social transformation, Cameroon can resist fragmentation and promote inclusivity, sustainability, and justice. This is especially important in Kom, where modernization pressures risk eroding communal bonds.

Equally, Mbih J. Tosam shows that *Ubuntu* values remain resilient in Kom even under modernity, offering a philosophical foundation for promoting the common good.[96] Equally, Benson O. Anofuechi and John S. Klaasen, in agreement with Okyere-Manu isolated a strong ethics of responsibility guaranteeing protection of identity in the face of technological advancements; absorbing *Ubuntu's* communal ethics. However, they commend the fact of the necessity of solidarity, reciprocity, and kinship obligations; acting as safeguards against fragmentation. Accordingly, they consider this as anchoring modernization in communal values and promoting justice as well as collective flourishing; thereby offering Africa a responsible model for negotiating modernity.[97]

Okyere-Manu's ethical reflections are highly relevant to contemporary Cameroon, demonstrated particularly within the Kom community; providing a framework for balancing modernization with cultural values. Technological transformation, though inevitable, must be critically managed to prevent the erosion of communal identity. By grounding innovation in African traditions of ethical responsibility, Cameroon gains a moral compass from Okyer-Manu and related authors, which directs progress toward strengthening society rather than fragmenting it. *Ubuntu* emerges here as the unifying principle, transforming modernization into a collective project that safeguards kinship, solidarity, and justice. The Kom society exemplifies this lived philosophy; illustrating that African values remain vital resources for navigating globalization. Therefore, Okyere-Manu's insights invite Cameroon through Kom cultural lens, to embrace technology with wisdom, embed ethics into innovation, and promote the common good through *Ubuntu*; ensuring that cultural identity and communal flourishing endure amid change.

General Conclusion

The central problem identified at the outset of this study was the growing disjunction between African socio-cultural values and the technological systems imposed upon them. While technology promises economic growth, improved healthcare, and enhanced education, it simultaneously undermines communal solidarity, erodes indigenous languages, and disrupts traditional practices. This dissertation has shown, through the worldview of Okyere-Manu, that such challenges cannot be addressed by uncritical adoption of foreign innovations without a deliberate integration of African ethical frameworks.[98]

Drawing on Okyere-Manu's articulation, *Ubuntu* understood as a lived ethics of solidarity, mutuality, and reverence for life emerges as a guiding moral compass for negotiating the challenges and promises of technological transformation.[99] By grounding innovation in communal responsibility, African societies can improve ethical responsibility in technology; ensuring that it serves dignity rather than commodification. The appraisal of Okyere-Manu's thought reveal the fact that African values are not static relics, but dynamic resources capable of guiding policy, education, and development in ways that preserve cultural identity while embracing progress.[100]

Equally, in application to the Cameroonian context, these ethical concerns highlight the urgent need for policymakers, educators, and communities to embed indigenous values into technological design and governance. Not only in Cameroon, but everywhere, governments must deliberately integrate African ethical traditions into drafted policies; educators should reform curricula to include African perspectives on technology, culture and ethics; communities must reclaim agency by actively shaping adoption of innovations; and innovators should embed indigenous knowledge systems into design processes. In this way, Africa can balance innovation with cultural preservation; ensuring that economic growth as projected through technological advancement does not compromise communal values or ecological harmony.

Thus, the reaffirmation of the central problem that socio-cultural values increasingly suffers erosion under the weight of technological advancement pressures calls for deliberate integration of African ethical values safeguard. By embedding perspectives of *Ubuntu* into policy frameworks, educational curricula, community engagement initiatives, technological design, and sustainable development strategies, Cameroon (in particular) and Africa (in general) can chart a trajectory of progress that is both technologically dynamic and socio-cultural culturally grounded. Yet, a critical challenge persists: how can African societies concretely institutionalize the ethics of communal responsibility within rapidly evolving technological systems while maintaining competitiveness in global innovation? This enables technological progress to be measured not only by its efficiency; but also by its impact on human dignity and communal well-being. In this way, societies would pursue technological advancements without sacrificing socio-cultural values. The comprehensive bibliography that ensues, constitute avenues of enhancing a reversal of degrading African socio-cultural values in the face of technological innovations.

Bibliography

Main Sources

Okyere-Manu B. - Antwi Joseph, *African Indigenous Knowledge System and Sustainable Development*, Research Gate, Pietermaritzburg, 2018.

Okyere-Manu Beatrice, *Intersecting African Indigenous Knowledge system and Western Knowledge System Moral Convergence and Divergence*, Cluster Publications Pietermaritzburg, 2018.

Okyere-Manu Beatrice, "African Ethical Values and Policy Making," in *Journal of African Philosophy*, Vol. 12, No. 2, 2018.

Okyere-Manu Beatrice, "*Ubuntu* and Development Practice in Africa," in *African Development Review*, Vol. 31, No. 1, 2019.

Okyere-Manu Beatrice, "Ethics, Environment, and Development in Africa", in *Journal of Sustainable Development in Africa*, Vol. 22, No. 3, 2020.

Okyere-Manu Beatrice, "The Place of The African Relational and Moral Theory of *Ubuntu* in the Global Artificial Intelligence And Big Data Discussion: Critical Reflections", in *Elgar Companion To Regulating Ai And Big Data In Emerging Economies*, Edward Elgar Publishing, Cheltenham, 2023.

Okyere-Manu Beatrice, African *Values, Ethics, and Technology Questions, Issues, and Approaches*, Springer Nature, Cham, 2021.

Okyere-Manu B.- E. Chitando, *The Palgrave Handbook Of Ubuntu, Inequality And Sustainable Development*, Springer Nature, Cham, 2024.

okyere-manu b. et al, "*Ubuntu* and Social Choice Theory: An Ethical Interrogation of Group Decision-Making from an African Perspective Bernardo" in *The Palgrave Handbook Of Ubuntu, Inequality And Sustainable Development*, E. Chitando-B. Okyere-Manu (eds.), Springer Nature, Gewerbestrasse, 2024.

Okyere-Manu Beatrice, "Responsible Technological Development and Ubuntu Ethics," in *the Palgrave Handbook of Ubuntu, Inequality and Sustainable Development*, Springer Nature, Cham, 2025.

Okyere-Manu Beatrice, "*Ubuntu* and Policy Implications for Sustainable Development," *in The Palgrave Handbook Of Ubuntu, Inequality And Sustainable Development*, Springer Nature, Cham, 2025.

Secondary Sources

Edozie K. Rita, *Pan Africa Rising: The Cultural Political Economy of Nigeria's Afri-Capitalism and South Africa's Ubuntu Business*, Springer, Cham, 2017.

Findlay Mark Et Al, *Elgar Companion To Regulating Ai And Big Data In Emerging Economies*, Edward Elgar Publishing Limited, Cheltenham, 2023.

Gyekye Kwame, *African Cultural Values: An Introduction*, Sankofa Publishing Company, Philadelphia, 1996.

Gyekye Kwame, *Tradition and Modernity: Philosophical Reflections on the African Experience*, Oxford University Press, Oxford, 1997.

Hountondji Paulin, *African Philosophy: Myth and Reality,* Indiana University Press, Bloomington, 1983.

Mbiti S. John, *African Religions and Philosophy*, Praeger, New York, 1969.

Ndasauka Yamikani *African Mind, Culture, and Technology Philosophical Perspective*, Springer Nature, Cham, 2024.

Nkwi G. Walter, *African Modernities and Mobilities: An Historical Ethnography of Kom, Cameroon*, Langaa, Bamenda, 2015.

Nussbaum C. Martha, *Creating Capabilities: The Human Development Approach*, Harvard University Press, Cambridge, 2011.

Ogude James, *Ubuntu and Personhood*, Africa World Press, Trenton, 2018.

Ramose M. Bernard, *African Philosophy through Ubuntu*, Mond Books, Harare, 1999.

Sen Amartya, *Development as Freedom*, Alfred A. Knopf, New York, 1999.

Tutu Desmond, *No Future without Forgiveness*, Random House, New York, 1999.

Wiredu Kwasi, *Conceptual Decolonization in African Philosophy: Four Essays*, Hope Publications, Ibadan, 1995.

Articles

Ajitoni D. Bukunmi, "*Ubuntu* and the Philosophy of Community in African Thought", in *Journal of African Social Sciences and Development*, Apas Africa, Vol. 2, No. 1, 2023.

Ajitoni D. Bukunmi, "*Ubuntu* and the Philosophy of Community in African Thought: An Exploration of Collective Identity and Social Harmony" in *Journal of African Social Sciences and Development*, Vol. 7, No. 3, 2024.

Anofuechi Benson – Klaasen John, "A critical analysis of *Ubuntu* as the nexus of identity development in present-day Africa," in *HTS Teologiese Studies/Theological Studies* vol.80, no. 1, 2024.

Camara S. Mohamed, "African Cultures and Representations in the Digital Era" in *African Media Space and Globalization*, Springer, Cham, no.1, 2023.

Edozie K. Rita, "*Ubuntu* and the Reconciliation of Technology with Justice in Africa," in *Journal of African Philosophy*, Vol.5, no.1, 2016.

Fagunwa Temitope, "*Ubuntu*: Revisiting an Endangered African Philosophy in Quest of A Pan-Africanist Revolutionary Ideology", in *Genealogy*, Vol. 3, Issue 2019.

Hofmeyr Benda "Reading Ubuntu Relationality through a Levinasian Lens," in *Acta Academica*, Vol. 57, No. 1, 2025.

Mbendera Richard, "*Ubuntu* Ethical Values And Africa's Quest For A Better Home" in *American Journal Of Humanities And Social Sciences Research*, Vol. 4, Issue 8, 2020.

Ndasauka Yamikani, "Rethinking Technology through African Philosophy," in *African Mind, Culture, and Technology*, Springer, Cham, no.1, 2024.

Nwosimiri Ovett, "African Cultural Values, Practices and Modern Technology," in *African Values, Ethics and Technology: Question Issues and Approaches,* Palgrave Macmillan, Cham, Vol. 1, no. 1, 2021.

Tosam J. Mbih, "Ɨwu-ɨ-Kom-ɨ-Twal: Kom Agrarian-Environmental Ethics", *in International Library of Environmental, Agricultural and Food Ethics,* Springer, Vol. 35, no.1, 2023.

---

[1] CF. J. S. MBITI, *African Religions and Philosophy,* Heinemann, London, 1969, 5.

[2] Beatrice Dedaa Okyere-Manu is a Ghanaian scholar specializing in Applied Ethics, with a strong focus on African values, technology, and socio-cultural transformation. She has authored and edited several influential works exploring how technological advancement intersects with African ethical frameworks. She is a Professor of Applied Ethics at the School of Religion, Philosophy, and Classics, University of KwaZulu-Natal, South Africa. She is a program director for Applied Ethics at University of KwaZulu-Natal. Education. She also holds a PhD in Applied Ethics from University of KwaZulu-Natal and a contemporary philosopher. Her main works include: *African Values, Ethics, and Technology: Questions, Issues, and Approaches Published: 2021, An African Perspective, Intersecting African Indigenous Knowledge*

Systems and Western Knowledge Systems: Moral Convergence and Divergence Published: 2018.

[3]Ubuntu Highlights communitarian spirit and African philosophical ethics of communal responsibility.

[4]Cf. B. Okyere-Manu, African *Values, Ethics, and Technology Questions, Issues, and Approaches*, Springer Nature, Cham, 2021, 105.

[5]Cf. K. Wiredu, *Cultural Universals and Particulars: An African Perspective*, Bloomington: Indiana University Press, 1996, 32

[6]Cf. B. Okyere-Manu, African *Values, Ethics, and Technology Questions, Issues, and Approaches*, Springer Nature, Cham, 2021, 1.

[7]Cf. M. Findlay et al, *Elgar Companion to Regulating AI and Big Data in Emerging Economies,* Edward Elgar Publishing Limited, Cheltenham, 2023, 103.

[8]Cf. M. Findlay et al, *Elgar Companion to Regulating AI and Big Data in Emerging Economies,* Edward Elgar Publishing Limited, Cheltenham, 2023, 99.

[9]B. Okyere-Manu- J. K. Antwi, *African Indigenous Knowledge System and Sustainable development,* Research Gate, Pietermaritzburg, 2018, 87.

[10]Cf. Ibid., 88.

[11]Cf. B. Okyere-Manu, *African Values, Ethics, and Technology Questions, Issues, and Approaches*, 2.

[12]Cf. Ibid., 2.

[13]Cf. B. Okyere-Manu, *African Values, Ethics, and Technology Questions, Issues, and Approaches,* 107.

[14]B. Okyere-Manu- J. K. Antwi, *African Indigenous Knowledge System and Sustainable development,* Research Gate, Pietermaritzburg, 2018, 14.

[15]Cf. Ibid., 14.

[16]Cf. M. Findlay et al, *Elgar Companion to Regulating AI and Big Data in Emerging Economies,* Edward Elgar Publishing Limited, Cheltenham, 2023, 103.

[17]Cf. M. Findlay et al, *Elgar Companion to Regulating AI and Big Data in Emerging Economies*, 106.

[18]Cf. B. Okyere-Manu, African *Values, Ethics, and Technology Questions, Issues, and Approaches*, 2021, 4.

[19]Cf. M. Findlay et al, *Elgar Companion to Regulating AI and Big Data in Emerging Economies*, Edward Elgar Publishing Limited, Cheltenham, 2023,105.

[20]Cf. B. Okyere-Manu, *African Values, Ethics, and Technology Questions, Issues, and Approaches*, 6.

[21]Cf. Ibid., 6.

[22]Cf. Ibid., 6.

[23]Cf. B. Okyere-Manu, *Intersecting African Indigenous Knowledge Systems and Western Knowledge Systems Moral Convergence and Divergence,* Cluster Publications, Pietermaritzburg, 2018, 14.

[24]B. Okyere-Manu, African Values Ethics and Technology Questions, Issues and Approaches, 3.

[25]Cf. B. Okyere-Manu, *Intersecting African Indigenous Knowledge Systems and Western Knowledge Systems Moral Convergence and Divergence,* 15.

[26]Cf. B. Okyere-Manu - E. Chitando, et al. (eds.), *The Palgrave Handbook of Ubuntu, Inequality and Sustainable Development,* Springer Nature, Cham, 2024, 32.

[27]Decolonizing technology means critically rethinking how technology is designed, used, and controlled so that it no longer reinforces colonial power structures, but instead respects and uplifts diverse cultures, knowledge systems, and communities.

[28]Cf. B. Okyere-Manu (ed.), *African Values, Ethics, and Technology: Questions, Issues, and Approaches,* Springer Nature, Cham, 2021, 13.

[29]Cf. Ibid., 5.

[30]Cf. B. Okyere-Manu et al, "Ubuntu and Social Choice Theory: An Ethical Interrogation of Group Decision-Making from an African Perspective Bernardo" *in The Palgrave Handbook of Ubuntu, Inequality and Sustainable Development,* E. Chitando-B. Okyere-Manu (eds.), Springer Nature, Gewerbestrasse, 2024, 685.

[31]Cf. B. Okyere-Manu, "African Ethical Values and Policy Making," *in Journal of African Philosophy*, vol. 12, no. 2, 2018, 45.

[32]Cf. B. Okyere-Manu et al, "Ubuntu and Social Choice Theory: An Ethical Interrogation of Group Decision-Making from an African Perspective Bernardo" *in The Palgrave Handbook of Ubuntu, Inequality and Sustainable Development,* E. Chitando-B. Okyere-Manu (eds.), Gewerbestrasse, Springer Nature, 2024, 687.

[33]Cf. B. Okyere-Manu, "Ubuntu and Development Practice in Africa," *in African Development Review*, vol. 31, no. 1, 2019, 78.

[34]Cf. B. Okyere-Manu, "Ethics, Environment, and Development in Africa", *in Journal of Sustainable Development in Africa*, vol. 22, no. 3, 2020, 102.

[35]Cf. B. Okyere-Manu (ed.), *African Values, Ethics, and Technology: Questions, Issues, and Approaches,* 2021, 2.

[36]Cf. Ibid., 6.

[37]Cf. B. Okyere-Manu, *African Values, Ethics, and Technology*: *Questions, Issues, and Approaches,* 15.

[38]Cf. Ibid., 62.

[39]Cf. Ibid., 101.

[40]Cf. B. Okyere-Manu et al, "Ubuntu and Social Choice Theory: An Ethical Interrogation of Group Decision-Making from an African Perspective Bernardo" in *The Palgrave Handbook of Ubuntu, Inequality and Sustainable Development,* E. Chitando-B. Okyere-Manu (eds.), Gewerbestrasse, Springer Nature, 2024, 690.

[41]Cf. B. Okyere-Manu, *African Values, Ethics, and Technology, Questions, Issues, and Approaches,* 3.

[42]Cf. B. Okyere-Manu, "The place of the African relational and moral theory of Ubuntu in the global artificial intelligence and big data discussion: critical reflections", in *Elgar Companion to Regulating AI and Big Data in Emerging Economies*, Edward Elgar Publishing, Cheltenham, 2023, 103.

[43]Cf. Y. Ndasauka, "Rethinking Technology through African Philosophy," in *African Mind, Culture, and Technology*, Springer, Cham, 2024, 104.

[44]Cf. K. Gyekye, *Tradition and Modernity: Philosophical Reflections on the African Experience*, Oxford University Press, Oxford, 1997, 40.

[45]Cf. B. Okyere-Manu, "The place of the African relational and moral theory of Ubuntu in the global artificial intelligence and big data discussion: critical reflections", 2023, 103.

[46]Cf. Y. Ndasauka, *African Mind, Culture, and Technology Philosophical Perspective*, Springer Nature, Cham, 2024, 1.

[47]Cf. B. D. Ajitoni *"Ubuntu and the Philosophy of Community in African Thought: An Exploration of Collective Identity and Social Harmony" in Journal of African Social Sciences and Development*, Vol. 7, No. 3, 2024, 1.

[48]Cf. R. Mbendera, *"Ubuntu Ethical Values and Africa's Quest for a Better Home"* in *American Journal of Humanities and Social Sciences Research*, Vol. 4, Issue 8, 2020, 177.

[49]Cf. B. Okyere-Manu, *African Values, Ethics, and Technology, Questions, Issues, and Approaches,* 108.

[50]Cf. Y. Ndasauka, *African Mind, Culture, and Technology: Philosophical Perspectives,* 15.

[51]Cf. K. Gyekye, *African Cultural Values: An Introduction*, Sankofa Publishing Company, Accra, 1996, 35.

[52]Cf. B. Okyere-Manu (ed.), *African Values, Ethics, and Technology: Questions, Issues, and Approaches,* 108.

[53]Cf. B. Okyere-Manu, *African Values, Ethics, and Technology, Questions, Issues, and Approaches,* 3.

[54]Cf. K. Gyekye, *African Cultural Values: An Introduction*, 37.

[55]Cf. M. B. Ramose, *African Philosophy Through Ubuntu*, Mond Books, Harare, 1999,49.

[56]Cf. K. Wiredu, *Conceptual Decolonization in African Philosophy: Four Essays*, Hope Publications, Ibadan 1995, 25.

[57]J. Ogude, *Ubuntu and Personhood*, Africa World Press, Trenton, 2018, 3.

[58]Cf. B. Hofmeyr "Reading Ubuntu relationality through a Levinasian lens," in *Acta Academica*, vol. 57, no. 1, 2025, 55.

[59]Cf. Y. Ndasauka, *African Mind, Culture, and Technology Philosophical Perspective*, 150.

[60]Cf. B. Okyere-Manu, *African Values, Ethics, and Technology, Questions, Issues, and Approaches*, 94.

[61]Cf. K. Gyekye, *Tradition and Modernity: Philosophical Reflections on the African Experience*, 42.

[62]Cf. A. Sen, *Development as Freedom*, Alfred A. Knopf, New York 1999, 18.

[63]Cf. P. J. Hountondji, *African Philosophy: Myth and Reality*, Indiana University Press, Bloomington 1996, 48.

[64]Cf. J. S. Mbiti, *African Religions and Philosophy*, Heinemann, London, 1969, 5.

[65]Cf. M. C. Nussbaum, *Creating Capabilities: The Human Development Approach*, Harvard University Press, Cambridge, 2011, 36.

[66]Cf. R. K. Edozie, *Pan Africa Rising: The Cultural Politics of African Renaissance*, Springer, Cham, 2017, 104.

[67]Cf. T. Fagunwa "*Ubuntu*: Revisiting an Endangered African Philosophy in Quest of a Pan-Africanist Revolutionary Ideology", in *Genealogy*, Vol. 3, Issue 2019, 45.

[68]Cf. B. D. Ajitoni "*Ubuntu* and the Philosophy of Community in African Thought", in *Journal of African Social Sciences and Development*, APAS Africa, Vol.2, No. 1, 2023, 15.

[69]Cf. K. Gyekye, *Tradition and Modernity: Philosophical Reflections on the African Experience*, 1997, 38.

[70]Cf. P. Hountondji, *African Philosophy: Myth and Reality*, 45.

[71]Cf. J. Mbiti, *African Religions and Philosophy*, 201.

[72]Cf. M. C. Nussbaum, *Creating Capabilities: The Human Development Approach*, Harvard University Press, Cambridge, 2011, 20.

[73]Cf. Chapter, 1.1.1, 8.

[74]Cf. K. Gyekye, *Tradition and Modernity: Philosophical Reflections on the African Experience*, 35.

[75]Cf. Supra, 1.4.2, 14.

[76]Cf. P. Hountondji, *African Philosophy: Myth and Reality*, Indiana University Press, Bloomington 1983, 49.

[77]Cf. A. Sen, *Development as Freedom* Oxford University Press, Oxford 1999, 31.

[78]Cf. J. S. Mbiti, *African Religions and Philosophy*, Heinemann, Oxford, 1990, 201.

[79]Cf. M. C. Nussbaum, *Creating Capabilities: The Human Development Approach*, Harvard University Press, Cambridge, 2011, 20.

[80]Cf. R. K. Edozie, *Pan Africa Rising: The Cultural Politics of African Renaissance*, Springer, Cham, 2017, 145.

[81]Cf. O. Nwosimiri, "African Cultural Values, Practices and Modern Technology," in *African Values, Ethics and technology: Question Issues and Approaches*, Palgrave Macmillan, Cham, 2021, 92.

[82]Cf. M. S. Camara "African Cultures and Representations in the Digital Era" in *African Media Space and Globalization*, Springer, Cham, 2023, 355.

[83]Cf. B. Okyere-Manu, *African Values, Ethics, and Technology: Questions, Issues, and Approaches*, 30.

[84]Cf. Ibid., 94.

[85]Cf. K. Gyekye, *Tradition and Modernity: Philosophical Reflections on the African Experience*, Oxford University Press, Oxford, 1997, 42.

[86]Cf. P. Hountondji, *African Philosophy: Myth and Reality*, Indiana University Press, Bloomington 1996, 53.

[87]Cf. R. K. Edozie, "*Ubuntu* and the Reconstitution of Community in Africa," in *Journal of African Political Economy*, 2017, 112.

[88]Cf. B. Okyere-Manu, *African Values, Ethics, and Technology: Questions, Issues, and Approaches*, 46.

[89]Cf. Cf. B. Okyere-Manu, *African Values, Ethics, and Technology: Questions, Issues, and Approaches*, 19.

[90]Cf. J. S. Mbiti, *African Religions and Philosophy*, Heinemann, Oxford, 1990,109.

[91]Cf. B. Okyere-Manu (ed.), *African Values, Ethics, and Technology:*, 2021, 5.

[92]Cf. B. Okyere-Manu (ed.), *African Values, Ethics, and Technology,* 2021, 47.

[93]Cf. M. J. Tosam, "Ɨwu-ɨ-Kom-ɨ-Twal: Kom Agrarian-Environmental Ethics", in *International Library of Environmental, Agricultural and Food Ethics*, Springer, 2023, 215.

[94]Cf. W. G. Nkwi, *African Modernities and Mobilities: An Historical Ethnography of Kom, Cameroon*, Langaa RPCIG, Bamenda, 2015, 65.

[95]Cf. W.G. Nkwi, *African Modernities and Mobilities: An Historical Ethnography of Kom, Cameroon*, Langaa RPCIG, Bamenda, 2015, 331.

[96]Cf. Mbih J. Tosam, "Ɨwu-ɨ-Kom-ɨ-Twal: Kom Agrarian-Environmental Ethics", in *International Library of Environmental, Agricultural and Food Ethics*, Springer, 2023, 230.

[97]Cf. B. O. Anofuechi and J. S. Klaasen, "A critical analysis of ubuntu as the nexus of identity development in present-day Africa," in *HTS Teologiese Studies/Theological Studies* vol.80, no. 1, 2024

[98]Cf. B. Okyere-Manu, *African Values, Ethics, and Technology: Questions, Issues, and Approaches*, 2021, 6.

[99]Cf. B. Okyere-Manu & E. Chitando (eds.), *The Palgrave Handbook of Ubuntu, Inequality and Sustainable Development*, 2024, 32.

[100]Cf. B. Okyere-Manu, *Intersecting African Indigenous Knowledge Systems and Western Knowledge Systems: Moral Convergence and Divergence*, 2018, 15.

Did you love *Degradation of African Socio-cultural Values in the Age of Technology : A Critical Engagement with Okyere Manu, Ubuntu and Communal Ethics*? Then you should read *From Human Connection to Human Replacement: The Ethical Risks of Social Robots and Companion AI*[1] by Emmanuel Ankia Gha!

[2]

What happens when the people who understand us best aren't people at all?Social robots and companion AI are no longer science fiction. They listen without judgment, remember everything, adapt to our moods, and are always available. They are being designed to care, to comfort, and to connect.

But at what cost?

In **_From Human Connection to Human Replacement_**, Emmanuel Ankia Gha pulls back the curtain on the rise of affective

1. https://books2read.com/u/3RRd1n

2. https://books2read.com/u/3RRd1n

computing and asks the question we've been avoiding: If we build machines to replace the warmth of human relationships, do we risk losing what makes us human?

This book explores the ethical risks hiding behind the promise of progress. From loneliness and emotional dependency to bias, manipulation, and the erosion of empathy, Gha examines how companion AI could reshape love, friendship, caregiving, and society itself.It's not a call to stop innovation. It's a call to keep it in check. Because progress without guardrails isn't progress. It's replacement.This book maps the ethical terrain of affective computing. Gha asks hard questions about what we gain and what we lose when we outsource connection to code.

Inside you'll explore:- The loneliness economy: How companion AI is marketed as the cure for isolation, and why it may deepen it instead- Emotional dependency: What happens to resilience, empathy, and human relationships when our primary confidant is an algorithm- Bias and manipulation: How data-driven "care" can reinforce stereotypes, exploit vulnerability, and nudge behavior without consent - Care and labor: The impact on nurses, teachers, therapists, and families as robots enter roles once defined by human presence- Identity and meaning: If a robot can love us back, do we still know what love is?

Gha doesn't argue for halting innovation. He argues for intentional innovation. The goal isn't to reject social robots, but to set boundaries, regulations, and cultural norms that keep humans at the center.Drawing on research in psychology, ethics, robotics, and real-world case studies, this book offers a framework for evaluating companion AI. Not just "can we build it?" but "should we?" and "under what conditions?"

Read more at https://smartworldpublishers.com.free.

About the Author

Akom Stephan Ngam is a Brother of the priesthood and a graduate researcher in Philosophy at the Catholic University of Cameroon, Bamenda (CATUC) - Bamenda. A dedicated scholar of African philosophy and ethics, his intellectual journey is rooted in both rigorous academic training and a profound vocational commitment to faith, service, and the search for truth.

Read more at https://smartworldpublishers.com.free.

About the Publisher

Smartworld Publishers is the publishing house of SWACEPH, dedicated to producing high-standard books that inform, inspire, and transform society. We publish research, educational materials, and stories in Education, Health, and Youth Leadership that address real needs across Africa and beyond. Every book we release is guided by quality, clarity, and purpose — to put knowledge into the hands of those who will use it to build stronger communities. Our mission is to provide an affordable and accessible platform for talented researchers. With Smartworld, publishing is made easy, affordable and impactful.

Read more at https://smartworldpublishers.com.free.

www.ingramcontent.com/pod-product-compliance
Lightning Source LLC
LaVergne TN
LVHW100200061126
851857LV00010B/343
* 9 7 9 8 2 3 3 7 5 7 9 3 8 *